DO YOU GET THE POINT?

by Dee Goldstein

Library of Congress Catalog Card Number:

ISBN-13: 978-0615891514
ISBN-10: 0615891519

Printed in the United States of America

This book is dedicated to:

My amazing husband Les ... who supports me always; my Mom and Dad ... who allowed me, while I was growing up, to be my own person; and our two wonderful children Troy and Dawn, and their respective super families of Lisa, Shelly, Shannon, and Mike, Max, and Davis ... who completely and forever smile and enjoy life with all of us!

CAN YOU

GUESS

THE ANSWERS???

just turn to the back section

THE POINTS!!!

... and find your path to a
totally enjoyable state of

"mathematical

delight"!!!

THIS BOOK COMES TO YOU IN TWO ENTERTAINING SECTIONS!

Part One: pages 1 – 100

Here, we have … clever, witty and humorous cartoons and graphics, with challenging mathematical overtones, that are designed to definitely put a big smile on your face, put a chuckle in your heart, and even raise your persona to an intensity of comical levity that can easily bring you to a very amusing "laugh out loud" moment of total enjoyment!

Part Two: pages i – xxxi

Here, we have … the answers for every page of this book's humor, with detailed explanations, and fully descriptive thought processes for each extremely special and individualized funny twist … all this, as one explores every facet of the mathematical inquisitiveness of one's mind!

ENJOY!!!

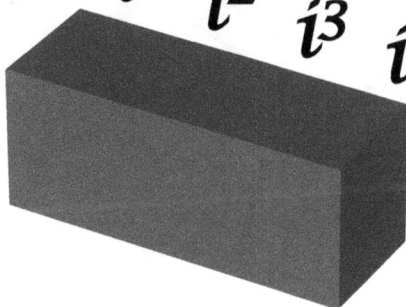

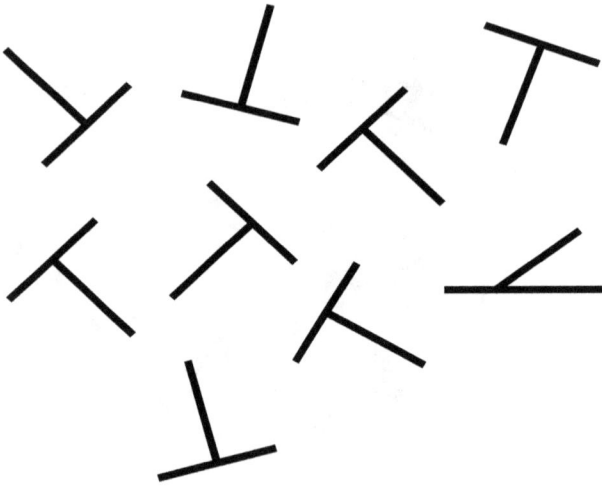

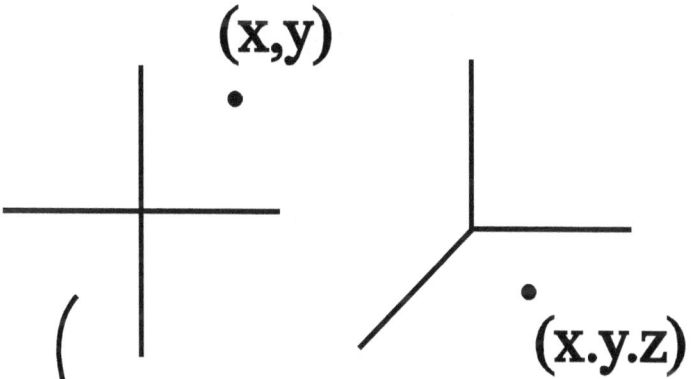

$$a + b = b + a$$

In "summation", and "commutatively", Bob is always great ... whether he is "coming or going"!

$$(1)(c) = c$$

$$(0)(d) = 0$$

3

Just "between"
you and me,
John really appears
to be quite "odd"!

2

4

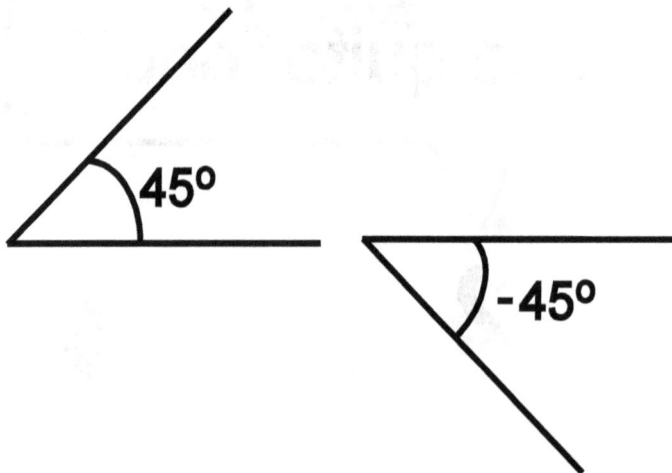

$$\sqrt{-1}$$

Anna,
her problem
is "purely
imaginary"!

{1,2,3,4,...n,...}

7

5

Jacob ...
it is only natural
that you can always
“count” on us!

.001

10 *100*

When we work out and get raised to higher positive integral powers, we get bigger and stronger ... poor Jerry just gets smaller and weaker!

x
d
z
c
b
y
a
Oh, to be just a bachelor again ... what "independence" I had!

$$Y = x^2 + 2x + 19$$

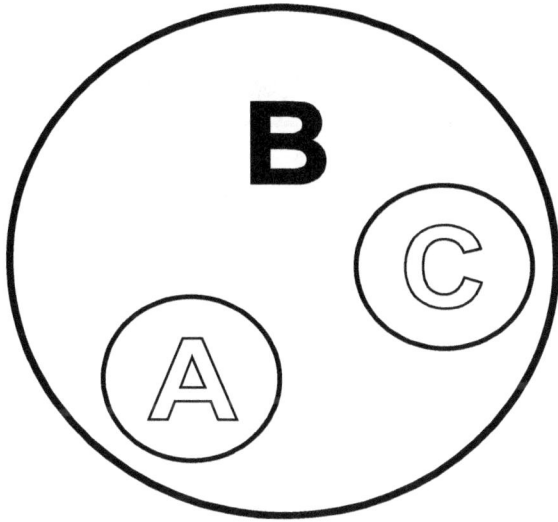

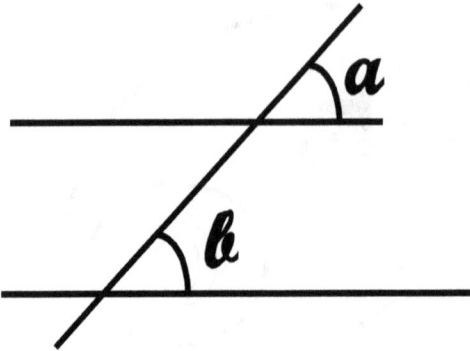

Kelly ...
all they do is
write and "correspond"
to each other
all day!

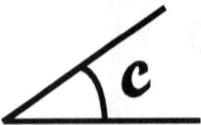

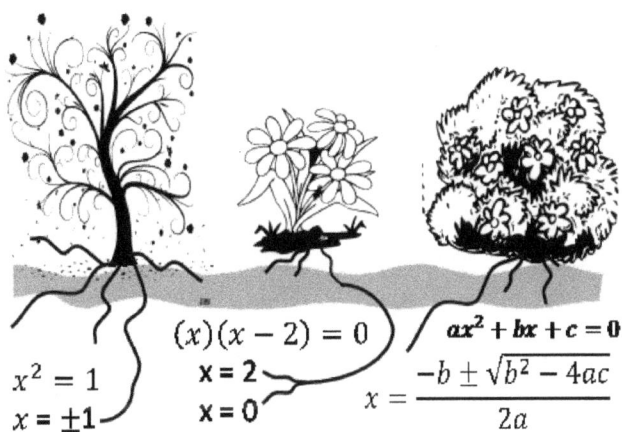

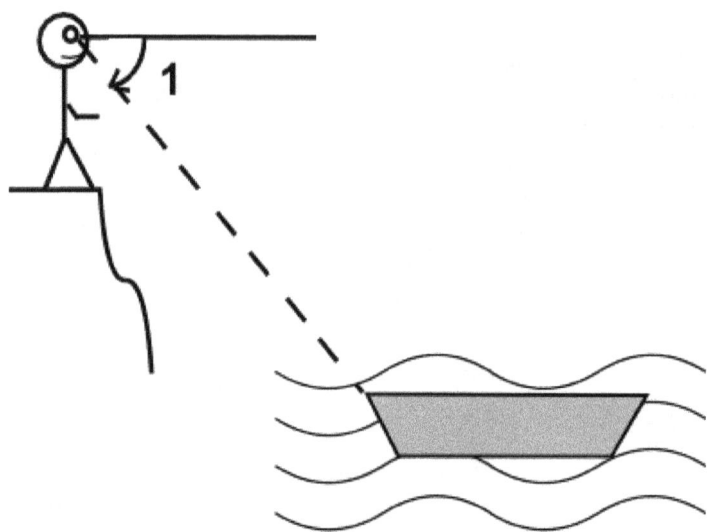

$\sin^2 x$

$1 - \cos^2 x$

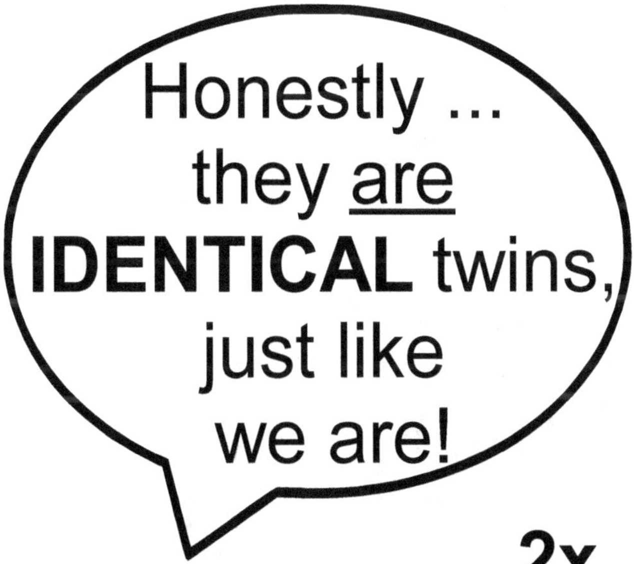

x

$\frac{2x}{2}$

$$a, a+d, a+2d$$

$$a, ar, ar^2, ar^3, ar^4$$

2   3

Dawn has progressed so much more rapidly in "geometry" than she has in "arithmetic"!

We are "obtuse" ...
but Taffy really
has "<u>a cute</u>"
sense of humor!

$$(x + 2)(2x) = y$$

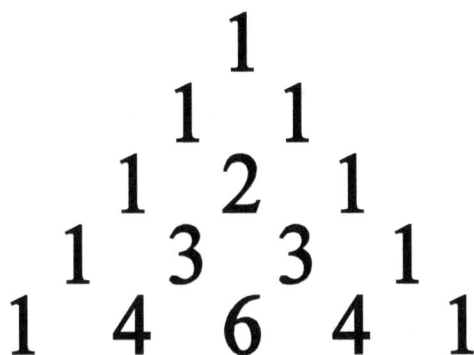

Yes, and Gayle owes all her popularity to her orthodontist!

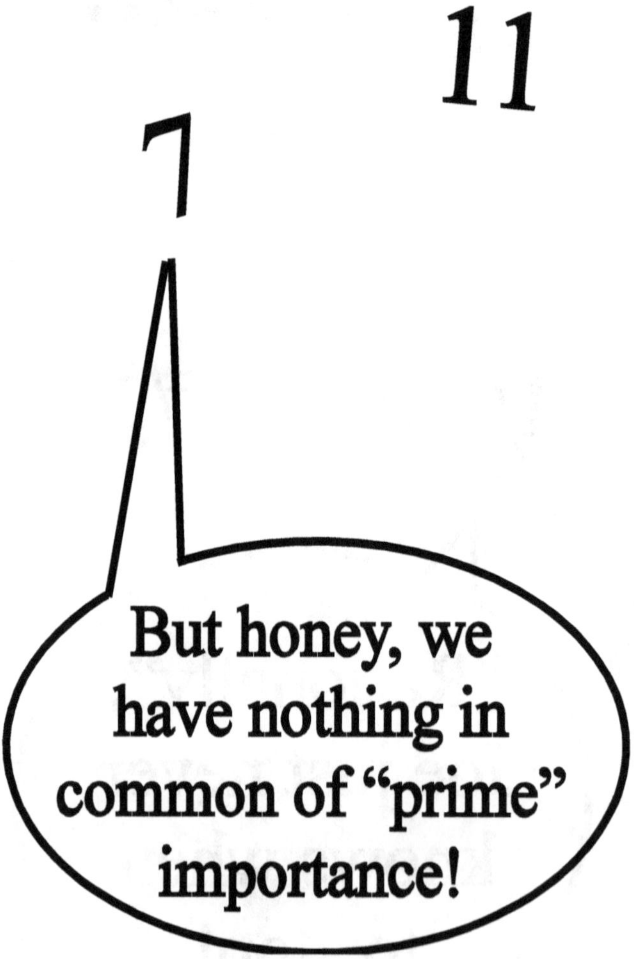
11
7
But honey, we have nothing in common of "prime" importance!

$$x^5y^4 + 7x^4y^2 + 8y^3 = z$$

$$a^2b^1 + a^1b^1 + b^2 = c$$

Misha,
I am sure that "z"
will get the job ...
after all, he does have
a "higher degree"!

8

3

If: "a → b"
and "b → c"
then ...

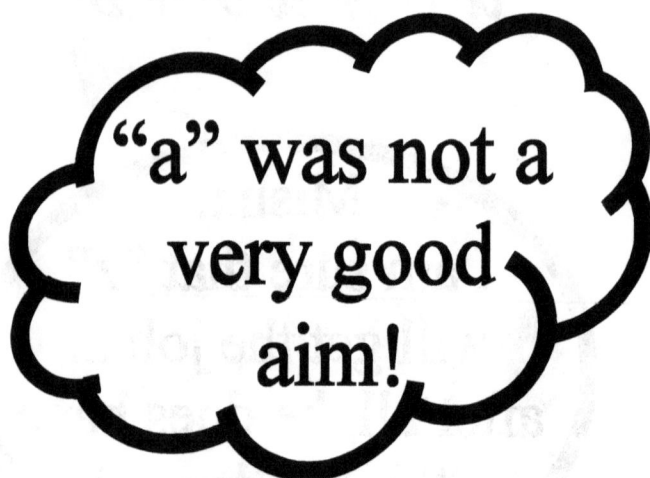

"a" was not a
very good
aim!

s

t

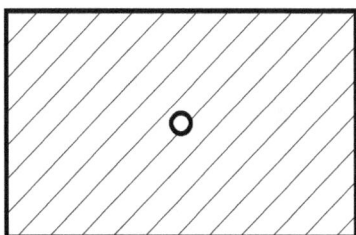

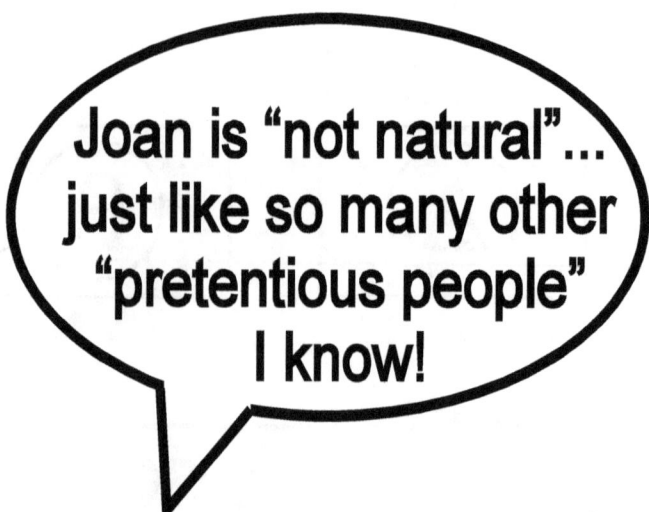

XIV

Bailey is obviously a "classics major"!

14

"fourteen"

$$\sqrt{2}$$

$$\sqrt{16}$$

$$\sqrt{9}$$

I am sorry, Sharon ... but your daughter is "not rational" and just does not have major human traits!

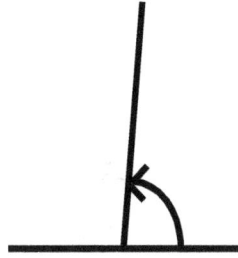

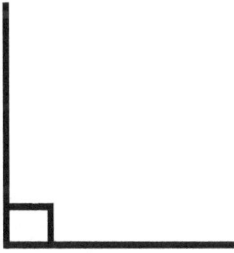

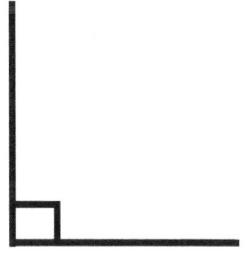

Jill is sometimes wrong and often inclined to "just miss being right"!

8

In "summing" up ...
my group had 43%
fewer cavities!

(x-5)

5	2	0	-40	-50
		10	50	50
	2	10	10	0

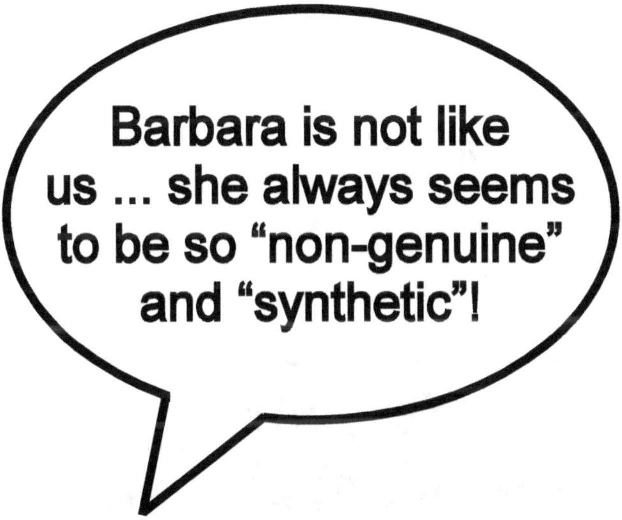

$\frac{5}{15}$ $6\overline{)13}$

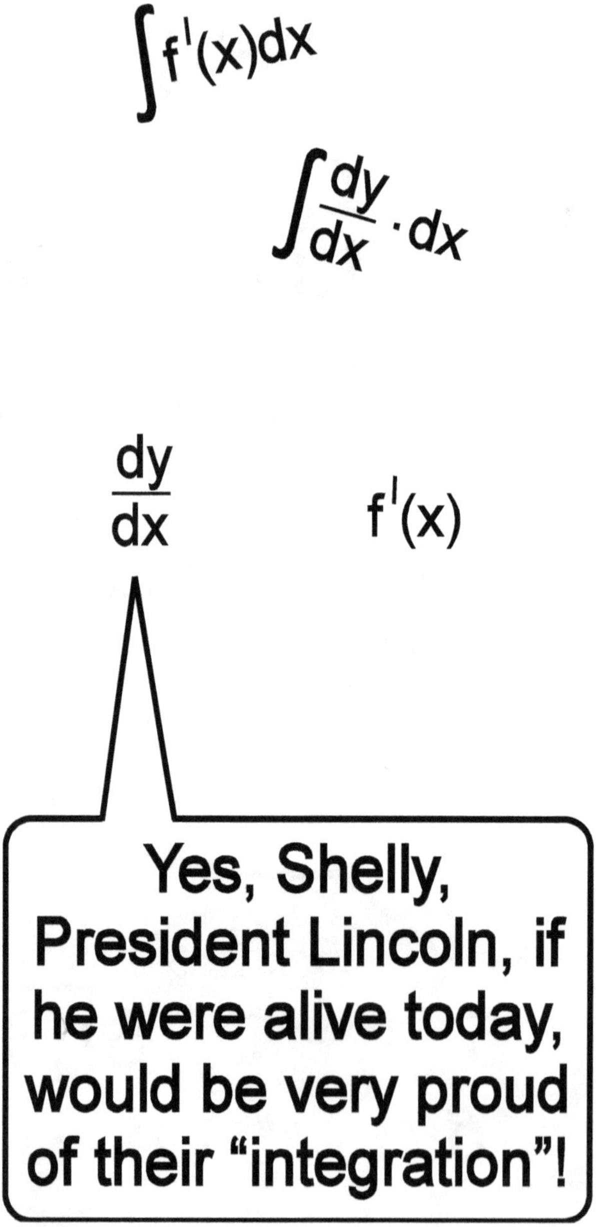
∫f'(x)dx
∫dy/dx · dx
dy/dx
f'(x)
Yes, Shelly, President Lincoln, if he were alive today, would be very proud of their "integration"!

{0}

{-1,-2,-3,-4,...}

{+1,+2,+3,+4,...)

-2

+1

Simon has always been such a "loner"!

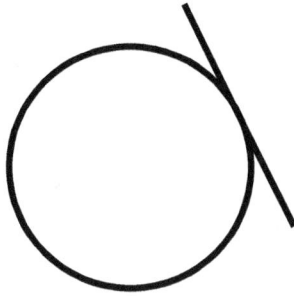

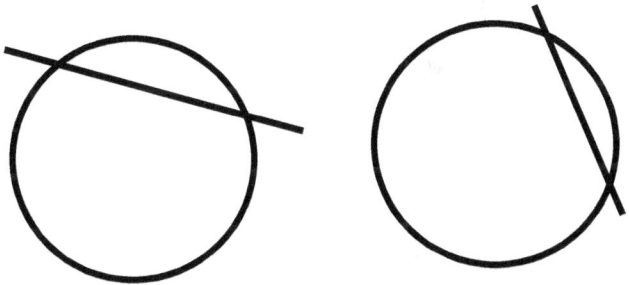

Since Jack is such a <u>tan</u> <u>gentl</u>eman, it is obvious that he must have spent his entire vacation in the Florida sun!

$\sqrt{1-y^2}$

$\sqrt{\frac{9}{121}}$

$\sqrt{\sin\theta}$

$\sqrt{4x}$

$\sqrt{5x-2}$

7

9

David,
I hope that our body
has antioxidants
to fight all those
free “radicals”!

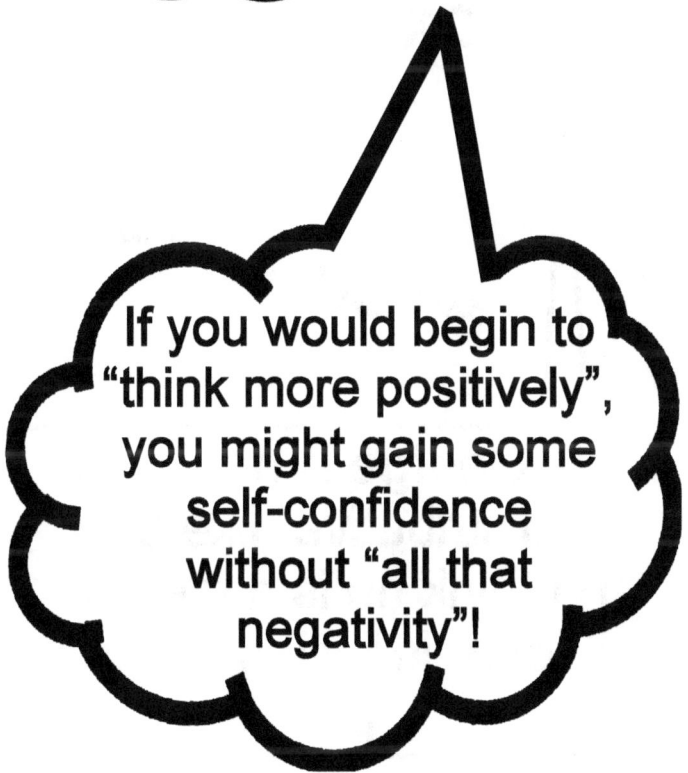

$$a(b+c+d)=ab+ac+ad$$

$$3x+5x = (3+5)x$$

$$a+ba = (1+b)a$$

George is not "right distributive" like us ... I think he is a member of a "left wing political party"!

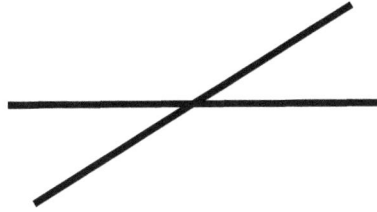

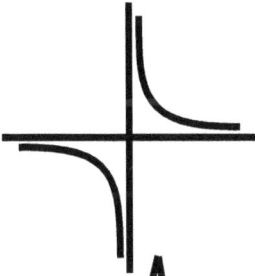

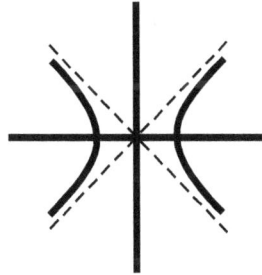

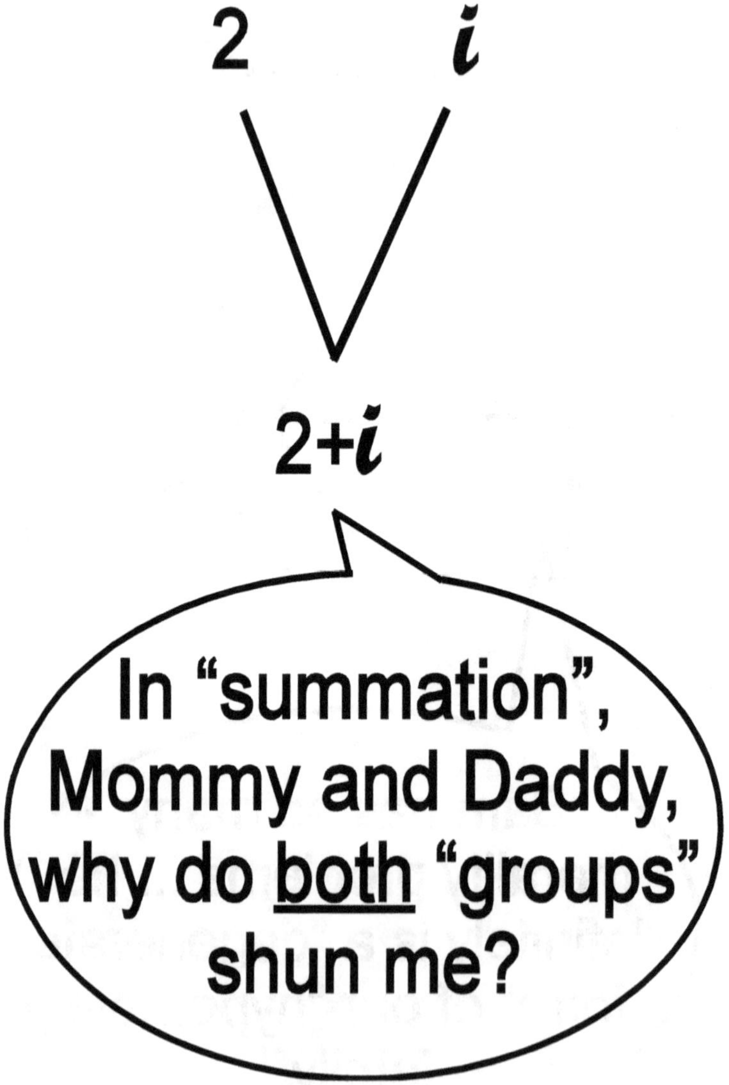
2
i
$2+i$
In "summation", Mommy and Daddy, why do both "groups" shun me?

$(x^2 + 4)$

$(x^2 + 2x)$

$(2x + 6)$

Mary never
will be really thin like
we are ... since she
never will be like us ...
"reducible in the
field of reals"!

$$\int \frac{dx}{1+x} \qquad \ln 7$$

$$e^x \qquad 2\ln y$$

$$\int \frac{dz}{z}$$

1

Linda, what wonderful "winter air" ... with all those "natural logs" burning in the fireplace!

2(mod 5)

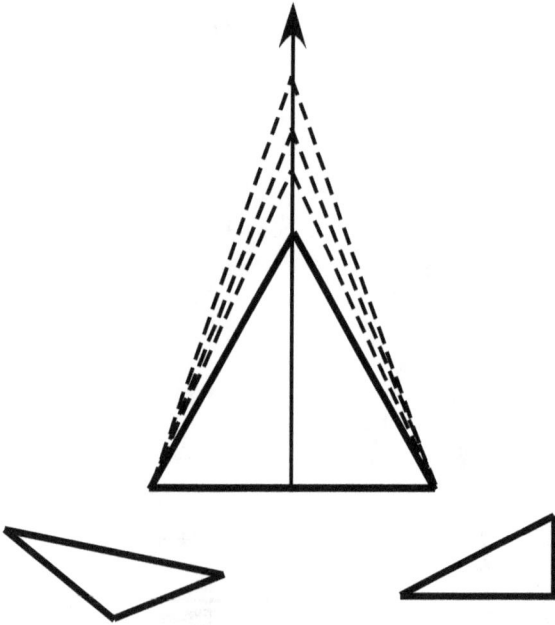

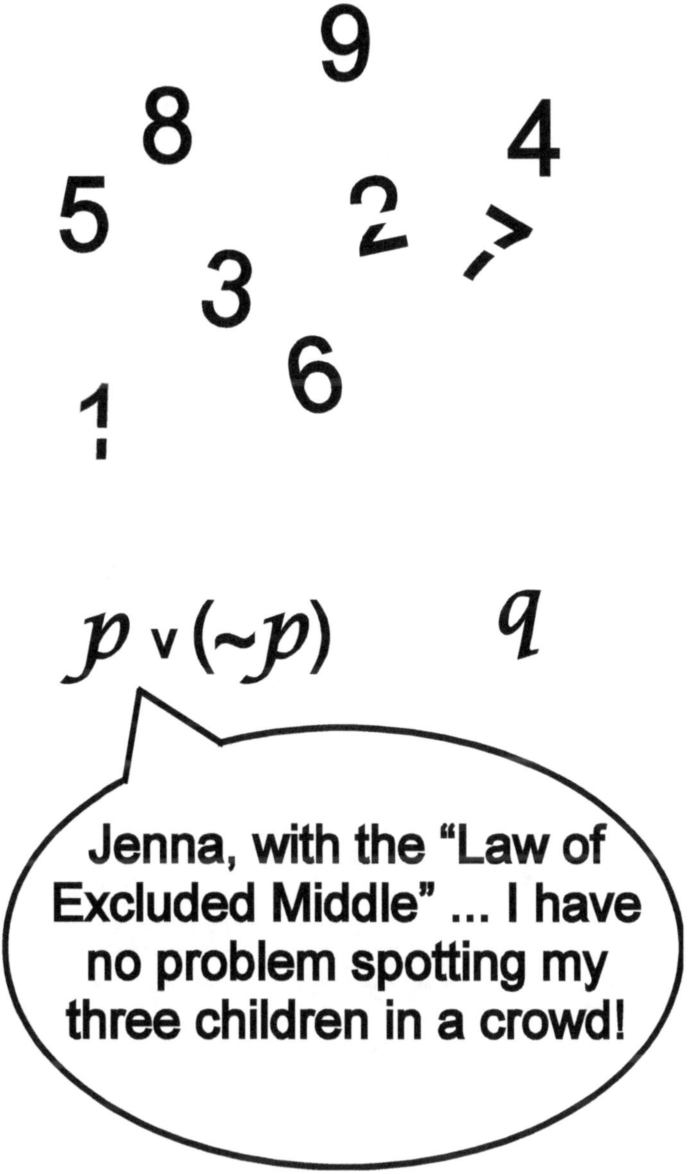
9
8
4
5
2
7
3
6
1
$p \vee (\sim p)$
q
Jenna, with the "Law of Excluded Middle" ... I have no problem spotting my three children in a crowd!

I am so dense and "obtuse" that, at first, I didn't even notice the strong resemblance and "similarity" that you two have!

$$d + 0\,i$$

$$a + b\,i \qquad c + b\,i$$
$$(b \neq 0)$$

Dear, because our "b" values are never zero, our "complex conjugates" never look like us ... but for Junior, it is his "identical" twin!

log 835.1

log 781.2

Les, as my brother ... we have both **inherited** the same "<u>positive characteristic of 2</u>"!

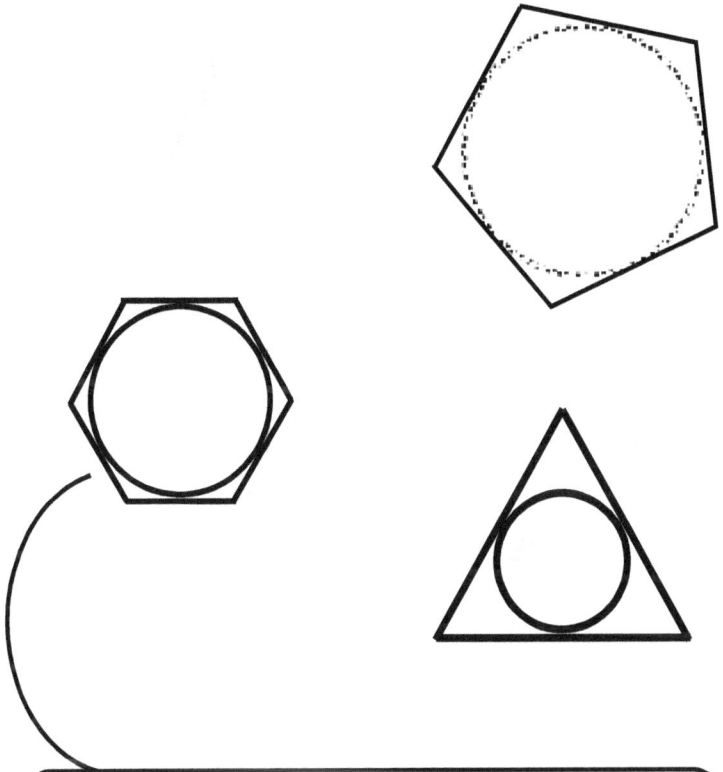

Don't mind Alex, he suffers "tangentially" from claustrophobia!

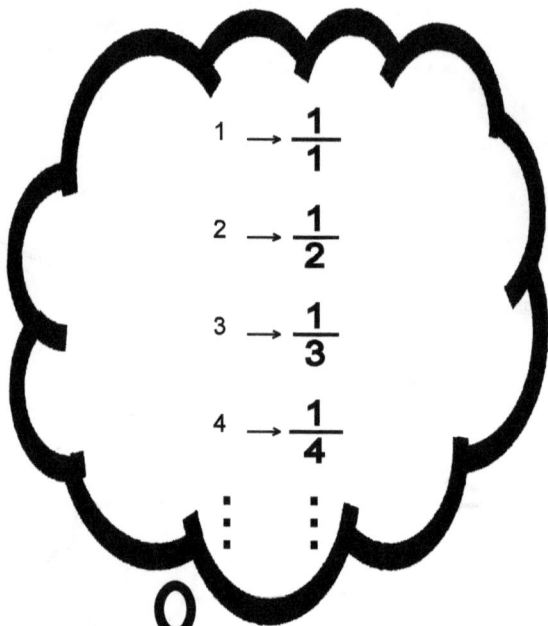

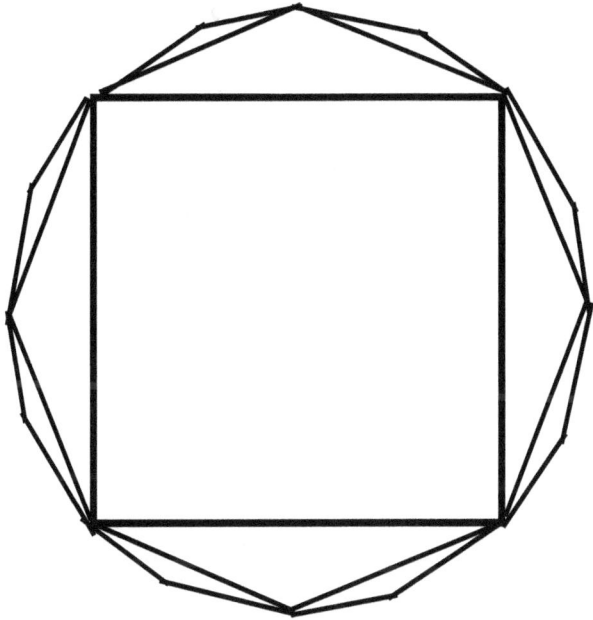

$$\frac{22}{7}$$

3.14

This "method of exhaustion" is really making Lisa tired so quickly!

8 2 6
14
16 4 12 28

5 9

In "summing" up ...
their group of even
numbers has always been
"closed under addition" to
us "oddities" ... and
always will be!

5 x 1

I think it is safe to assume that our "product children" will look more like me than you!

$$1+2+3+4+5 = 15$$

$$\sum_{i=1}^{6} i \qquad \sum_{i=1}^{5} i$$

Troy, how long has your twin been away from "Greece"?

$$\int \sin x \cos x \, dx$$

$$\int_0^{\pi} \sin\theta \, d\theta \qquad \int_{-4}^{-1} 3t \, dt$$

Julia is always "indefinite" ... her indecision always dominates!

$$\log_{10} 100 = 2$$

$$100 = \text{antilog}_{10}\ 2$$

5   9

For any idea, there is always someone to politically fight "against" it!

π

7

11

Is Mike really supernatural ... or just "transcendental"?

$$b^2 - 4ac = 0$$

 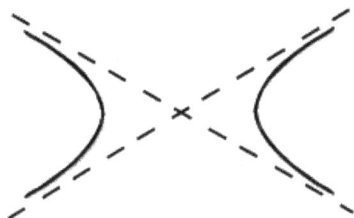

Chuck's family has no "discrimination" or prejudice in one direction or the other ... they merely have "discriminants" equal to zero ... as compared to our positives and negatives!

220  284

smack

5

2

Mark and Beth are so "amicable" and friendly ... all because the sum of each of their proper divisors equals the other!

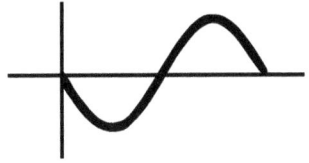

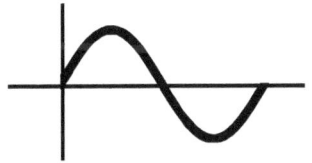

$$\frac{d^2y}{dx^2} = 0$$

$$\frac{d^2y}{dx^2} = -1$$

$$\frac{d^2y}{dx^2} = 1$$

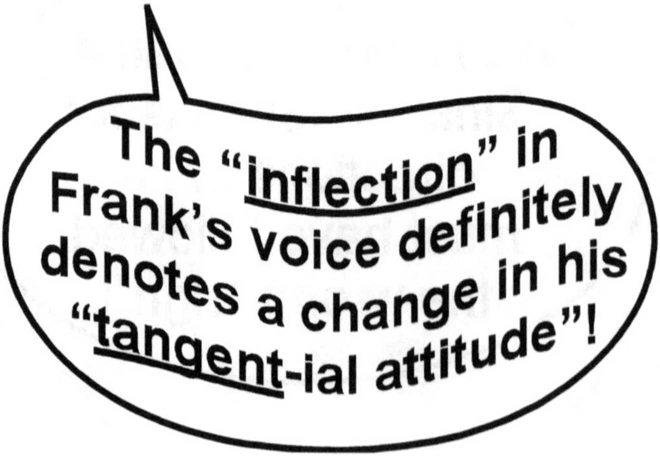

$$Z = Z$$

By "DUALITY", if we interchange the lines and points, the Western Gunslinger is as good as the Zorro Swordsman!

"Unknown cubed 1,
Unknown squared 13,
Unknown 5"

$$X^3 + 13X^2 + 5X$$

Tammy may be more
"symbolic", but
in "syncopated algebraic"
terms ... I have way
more rhythm!

$$\frac{du}{ds} = \text{grad } u \, \frac{\vec{v}}{|\bar{v}|}$$

$$\frac{dy}{dx} \qquad \frac{dr}{dt}$$

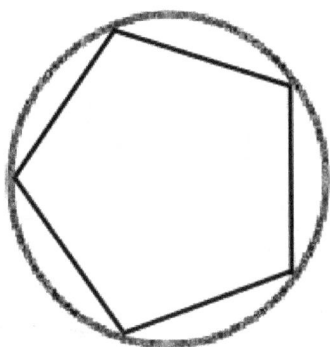

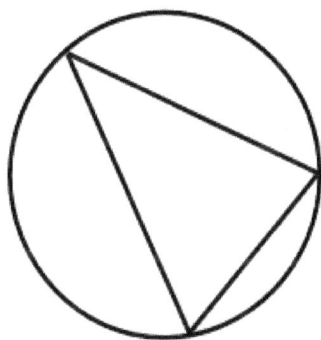

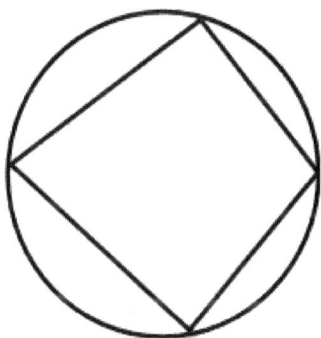

Karen is always so "ticklish"!

2   9

It looks like
a "cycloid" hit
Steven's room!

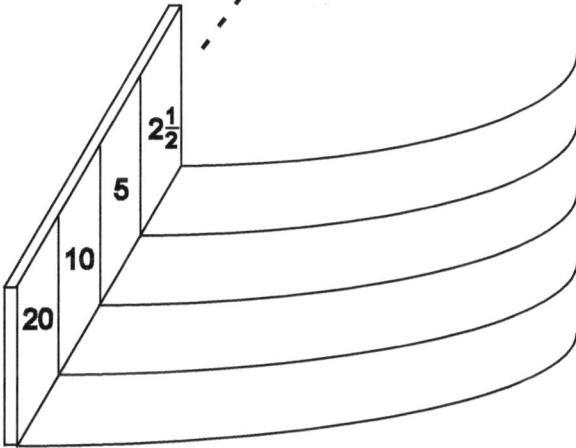

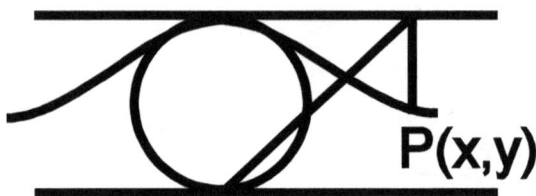

eq (1) $\quad x^3 - 3x^2 + 2x - 3 = 0$

height of eq (1) = |1| + |-3| + |2| + |-3| + (3 - 1) = 11

eq (2) $\quad 3x^3 - 9x^2 + 6x - 9 = 0$

height of eq (2) = |3| + |-9| + |6| + |-9| + (3 - 1) = 29

4

10

Thomas has gotten so "tall" ... one can hardly recognize him!

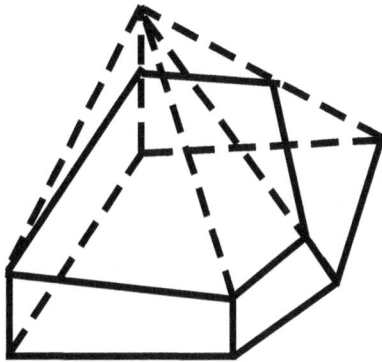

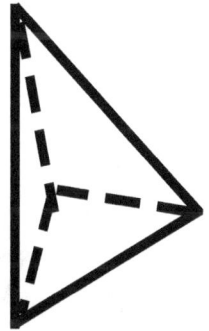

No ... Barry
is not in the
elephant family, he is
just one of our
"truncated" relatives!

$$f(x,y,c) = 0$$

$$f(s,t,c) = 0$$

$$F(s,t,t') = 0$$

$$F(x,y,y') = 0$$

Dani, what
kind of sophistication
can one expect from those
“Primitive Equational Civilizations”
as compared to our “Advanced
Differential Society”?

$$y = 2x^2 + cx + d$$

$$\text{where } \frac{d^2y}{dx^2} = +4$$

$$y = -2x^2 + cx + d$$

I never have to worry about your problem of "mid-day slump" ... because my "second derivative" is always "negative"!

$$\begin{pmatrix} 5 & x+iy \\ x-iy & 4 \end{pmatrix}$$

$$\begin{pmatrix} a & b & c \\ -b & 0 & 1 \\ -c & 4 & 2 \end{pmatrix} \quad \begin{pmatrix} r & s & t \\ -t & 1 & s \\ r & t & r \end{pmatrix}$$

Harold is a "Hermitian Matrix" ... and has always enjoyed "being by himself"!

$$f(t) = ae^{bt}$$

10 11

Kathleen, our "growth curve" equation is here to stay as long as "bacteria" never consider birth-control!

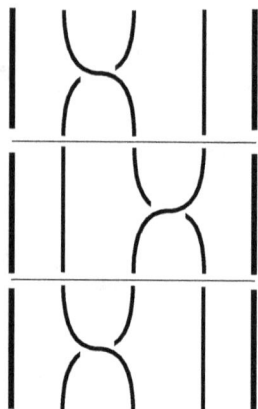

$$\sigma_i \sigma_{i+1}^{-1} \sigma_i$$

8   9

Alicia loves her "braids" and pigtails ... she never wants to grow up!

$$
\begin{array}{r}
p \rightarrow q \\
q \rightarrow r \\
\hline
\therefore p \rightarrow r
\end{array}
$$

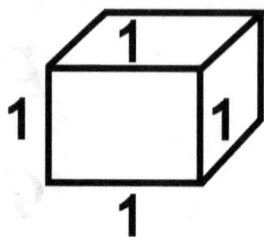

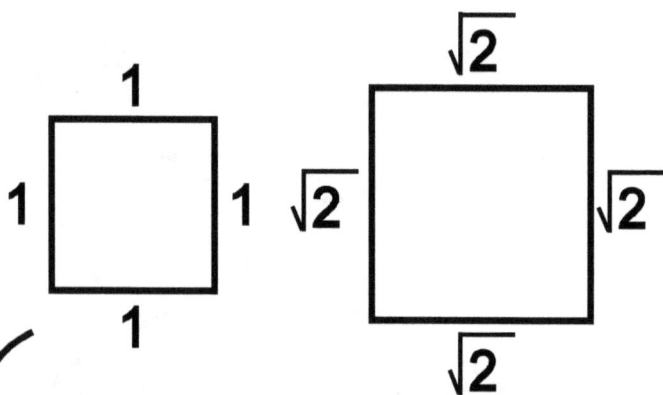

Considering the impossibility ... with only the use of Euclidean Tools ... that poor "lonely cube" will never have a "duplicate partner" like you are to me!

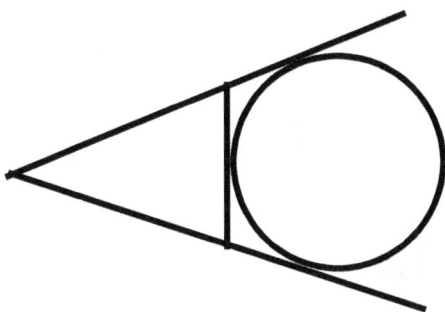

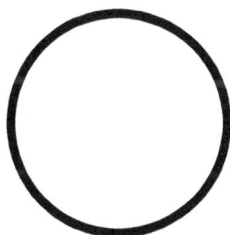

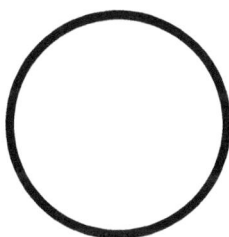

He used to be
one of us ... but
now all his perfect
tangents make
him, by definition,
an "ex-circle"!

~~1~~ 2 3 ~~4~~ 5

~~6~~ 7 ~~8~~ ~~9~~ ~~10~~

11 ~~12~~ 13 ~~14~~ ~~15~~ ...

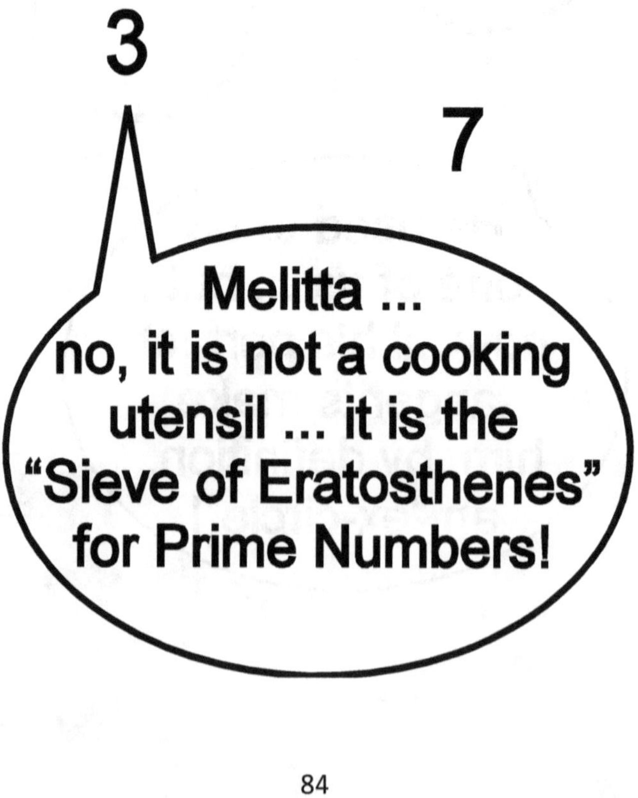

$$\begin{pmatrix} 1 & 2 & 5 \cdot y_1 \\ 1 & 0 & 1 \cdot y_2 \\ 5 & 0 & 0 \cdot y_3 \end{pmatrix}$$

$$\begin{pmatrix} 1 & 2 & 5 \\ 1 & 0 & 1 \\ 5 & 0 & 0 \end{pmatrix} \qquad \begin{pmatrix} 3 & 7 & 4 \\ 2 & 9 & 6 \\ 1 & 5 & 8 \end{pmatrix}$$

Even after the "augmentation" operation, Yetta is still my "identical" twin!

$$(x^2 + y^2)^3 = 4x^2y^2$$

13   7

Since I have found "Eli, the Four-Leaf Clover" ... my luck has totally changed for the better!

$$\{\,1,\,2,\,3,\,...\}$$
$$\updownarrow\ \updownarrow\ \updownarrow$$
$$\{\,5,\,6,\,7,\,...\}$$

10

2

The opposite approach is the "ordinal approach", not the "sacrilegious approach"!

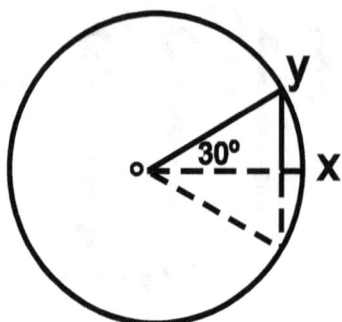

5

4

9

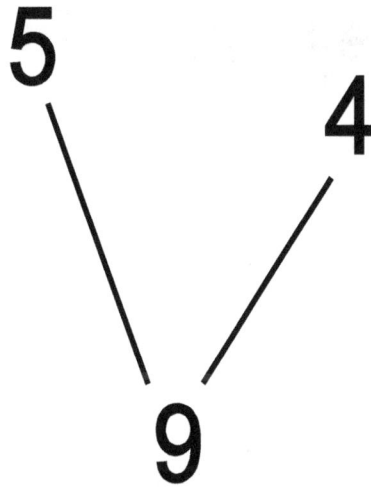

Pythagoreans consider "odd numbers" to be "masculine" ... thus, I think we have just given birth to a baby boy!

$$|\text{monkey}|\,|\text{lion}|\,\vec{n}\sin\theta$$

monkey

lion

With a "cross-product" like that .... what were we trying to "breed"?

1, 1, 2, 3, 5, 8, ..., x, y, x+y, ...

11   3

Yes ... by viewing this
"Fibonacci Sequence"
and knowing the close
connection that it parallels
to "Rabbit Reproduction" ...
you now know why I have
just invested in a
Carrot Farm!

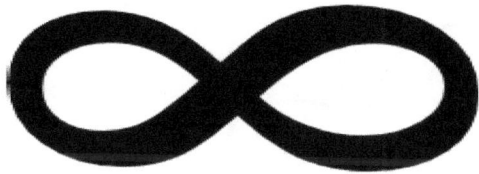

$$r^2 = a^2\cos2\theta$$
$$= \text{``}\infty\text{''}$$
$$= \text{infinity???}$$

$$y = a \cosh\left(\frac{x}{a}\right)$$

Now that he's an octopus (his <u>OCTO</u> from eight, and his <u>PUS</u> - sy - cat from our <u>CAT</u> - enary curves) ... he considers himself too good to "hang around" with us anymore!

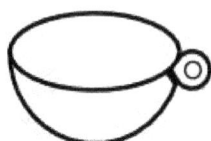

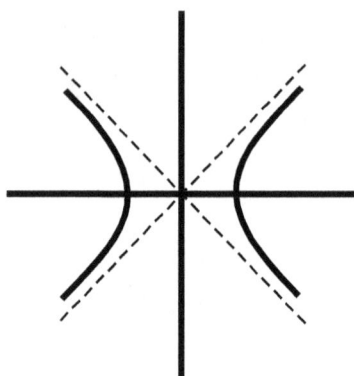

$$y = e^{-.7t}\cos 4t$$

$$y = e^{-.2t}\cos 2\pi t$$

With "DAMPED HARMONIC MOTION" ... it is too HUMID for me to move!

(1,1,1)

(2,2,2)

(5,5,5)

(3,3,3)

(9,9.9)

(0,0,0)

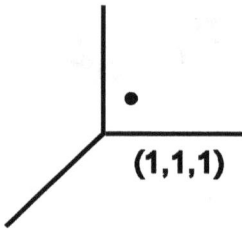

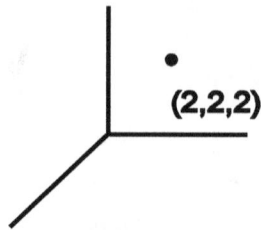

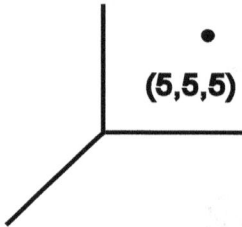

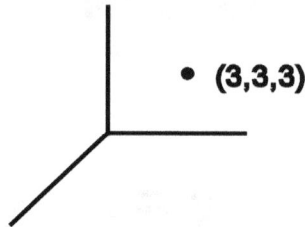

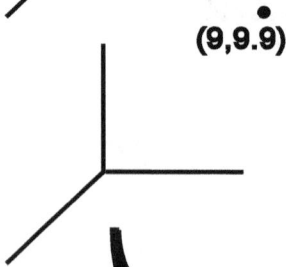

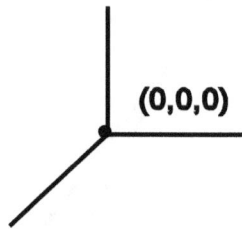

Tune in tomorrow
for another session
of "Exercising
with Isometrics" ...
or ... "let's all be
equidistant from
our three axes"!!!

THE

POINTS

Page	Point
1	$i = \sqrt{-1}$, $i^2 = -1$, $i^3 = -\sqrt{-1}$, $i^4 = 1$. The first and third are imaginary and the second is "negative one". Therefore, only the last one is both positive and equal to one … thus, it is "the real number 1".
2	Something that is normal is at right angles or perpendicular. Here, only one pair of lines is not perpendicular to each other.
3	Herman is just a point who has 2 coordinates that are depicted as two blanks on either side of the commas within the parentheses … thus, he has no specific coordinates to place him in a specific location with respect to the "axis system".
4	The commutative property of addition confirms that "order" is not important when one calculates the results of the "sum" of two numbers.

5	Three is the <u>odd number between</u> two and four.
6	$45^0 + (-45)^0 = 0$, which is "nothing". However, $45^0 + (+45)^0 = 90^0$, which is a "right" angle.
7	$\sqrt{-1}$ is a "<u>pure imaginary</u>" number.
8	The <u>natural numbers</u> are the <u>counting numbers</u>.
9	When numbers >1 (e.g. 10 or 100) get raised to increasing positive integral powers, they get larger and larger. However, when .001 (0<a number<1) gets raised to increasing positive integral powers, this value gets smaller and smaller, while still remaining between 0 and 1.
10	The "bold y", in the equation below, is the "<u>dependent</u> variable", in contrast to the "shadowed y" above which is alone as an "<u>independent</u> variable".
11	A and C are both subsets of B. Thus, when we consider the union of "A U B U C", the result is still

	equivalent to only B.
12	"a" and "b" are both <u>corresponding</u> angles of parallel lines ... thus, this capitalizes on "correspondence by writing letters" to one another.
13	The equations are beneath the ground, and the roots of the flowers, plants, and trees are actually mathematical <u>roots of these respective equations</u>.
14	Angle 1, lying on the couch, is the <u>"angle of depression" up above</u>.
15	If we consider the values of "Sin^2x" and "$1 - Cos^2x$" they are equal in value. If we look at "$Sin^2x = 1 - Cos^2x$", this is a very common "mathematical identity" used to solve many trigonometric identity problems.
16	There are more terms pictured above in the <u>geometric progression</u> than in the <u>arithmetic progression</u>.
17	Pictured are two <u>obtuse</u> angles talking about an <u>acute</u> angle above.

18	An "argument" is the <u>independent variable</u> upon whose value that of another quantity depends. Here, "$2x$" and "$x+2$" are the arguments, and y is the result of the <u>multiplication</u> of these and is thus called the <u>product</u>.
19	"Pascal's Triangle" represents the triangular array of the binomial coefficients. One can note that this pattern is identical whether it is read from "left to right" or read from "right to left" – here, we can actually use the physical direction of the Arabic, Farsi, or Hebrew Languages that read from right to left … and one can easily use these same identical coefficients in that direction as well.
20	What is to be noticed is the <u>perfect alignment of the number seven's teeth</u> in the crowd, as compared to the crooked teeth belonging to the <u>other two number sevens</u> that are talking to each other.

21	$.333\overline{3} \ldots$ is a rational number that is a non-terminating and repeating decimal. Here the "bar over the 3" denotes that this 3 keeps repeating forever. This is in contrast to the "1/5" and "1/2" that have a "finite decimal expansion". They are all rational numbers. Joe is rational because any real number which has an eventually periodic decimal expansion is considered to be a rational number. Also, one generally defines a rational number to be that which can be expressed as "a/b" where a and b are integers and b ≠ 0. Here, all three numbers belong to the set of rational numbers.
22	The definition of a prime number is "a whole number that is greater than 1 and also can only be divided evenly by 1 or itself". Thus, the number "7" and the number "11" have <u>no common prime factors</u>.

23	The equation of "z" is of <u>degree 5+4=9</u>, while the equation of "c" is only of <u>degree 2+1=3</u>.
24	Since this is read "if a implies b" and "b implies c", it would then normally follow logically, by "Aristotle's Law of Syllogism" or the "transitivity property of implication", that the typical next response would be ... "then a implies c". However, since this "implies" translates by mathematical symbols into arrows ... this capitalizes on the concept of an Archer who shoots a "Bow and Arrow".
25	Sue is very small compared to her universe. The <u>complement of a set</u> is everything in the universe that is not in the set. When one always gives "<u>complimentary remarks</u>", they are often considered to be exaggeratedly sweet or cloyingly agreeable. This plays on the fact that, although these words are spelled differently, they are

	pronounced identically.
26	Here we have <u>two natural numbers</u> that are talking about $\frac{2n+1}{2n}$, which is a number that is <u>not natural</u>.
27	**XIV** is another expression for those same number "fourteens" … however, this is classified as a <u>Roman</u> numeral. "Classics" is the study of the "ancient <u>Greek</u> and <u>Roman</u> cultures".
28	$\sqrt{2}$ is irrational … this is the opposite of the human trait of rationality. Here we have two "square roots" of perfect squares that are thus rational, talking about the square root of a whole number "not able to be expressed as the product of perfect squares" which is thus irrational.
29	The <u>inclination</u> of the angle shown is <u>slightly less than 90^0</u>. Since <u>90^0</u> is defined as a <u>right angle</u> and Jill's angle is only a few degrees less than this … she just misses being a "<u>right</u> angle".
30	These "8's" can be considered to be

	members of a mathematical "group" of even numbers which are closed under the binary operation of addition ... or "summing" up! This drawing exemplifies a familiar commercial for Gleem toothpaste that had always depicted 2 separate groups of individuals ... with one group always stating that "our group had a "certain percentage of fewer cavities than the other group"! Here, the dotted lines on the second group of 8's exemplify the "cavities" in the teeth of the latter group!
31	Barbara is synthetic division ... as contrasted with the other two ways of showing a more typical and common representation of actual division.
32	In mathematics, the opposite of integration is differentiation. While in Lincoln's era ... the opposite of "integration" was segregation. Here, two "differentiations" discuss those

	"integrations" pictured above.
33	Pictured are the set of <u>positive numbers</u>, the set of <u>negative numbers</u>, and <u>zero</u> is all by himself.
34	By the Venn diagram shown, the set <u>intersection</u> of A and B is the <u>null</u> set, or the <u>empty</u> set.
35	This capitalizes on the concept of a <u>gent</u>leman getting <u>tan</u> from the sun. Here we use the first part of the word <u>gent</u>leman, after that word <u>tan</u>, in order to get the common mathematical term of "<u>tangent to a circle</u>". Mathematically, Jack is a <u>tangent</u> to a circle, while the other two are both chords within a circle.
36	Pictured are <u>radicals</u> ... which, in mathematical symbols, are square root signs. Free radicals can cause damage within our body. Scientists have found these free radicals to be able to be successfully combatted with the use of antioxidants.

37	Pictured are the concepts of <u>positive and negative infinity</u>.
38	Two examples of the law of the <u>right-hand distribution</u> are talking about the <u>left-hand distributive law</u> at the top of the page. In politics, the term "left wing" generally refers to more "liberal" or "progressive" views.
39	Two common hyperbolas are discussing the <u>degenerate form</u> of the <u>family of hyperbolas</u> which is the intersection of two lines. The common mathematical use of the term "degenerate" is here compared with the English use of the same word to usually describe a person who is frowned upon by society for a below standard moral character.
40	A real number and a pure imaginary number combine by addition to form a <u>mixed</u> imaginary, or complex number. The pure reals and the pure imaginaries each constitute a

	mathematical <u>group</u>, which, under the operation of addition ("in summation"), are both closed … and <u>of which the resulting sum that is depicted cannot belong</u>.
41	Two terms which are <u>reducible</u> in the field of <u>reals</u> are discussing a term which is "simply unfactorable" … thus <u>irreducible in the field of reals</u>.
42	Everything pictured above represents <u>natural log</u> functions. Logs are often used in the fireplace during those cold winter months.
43	5 is equivalent to <u>"0"</u> when it is "converted" into mod 5 … and we think of the number "0" as "nothing". Here, one also often refers to "converting religions" in order to have a marriage ceremony in an identical faith.
44	Sarah is going to a <u>higher altitude</u> mathematically as the perpendicular to the base of the triangle increases …

	however, when exercising at higher altitudes above sea level, a higher altitude can often negatively affect one's endurance.
45	p v (~p) is the "<u>Law of Excluded Middle</u>". This asserts that, for every "proposition p" ... either "p is true" or "p's negation is true". This is one of the "Three Fundamental Laws of Logic".
46	The triangle that is talking is <u>dull</u> because he is an <u>obtuse</u> triangle. The other two triangles are <u>marked</u> to show that "<u>angle-angle-angle</u>" equals "<u>angle-angle-angle</u>" which is a condition for those two triangles to be termed "<u>similar triangles</u>".
47	Two lines, in two dimensional Euclidean space, are said to be parallel if they are always the <u>same distance apart</u> and ... thus, they will <u>never meet</u>.
48	The "two complex number parents

	(whose b values are never zero)" are talking to each other and commenting about the obvious fact that their son's imaginary component is zero … thus making Junior's complex conjugate of "d-oi" identical to his personal innate value of "d+oi". "Complex conjugates" are defined as a pair of complex numbers, both having the same real part, but with imaginary parts of equal magnitude and opposite signs.
49	Two non-common fractions are talking about the <u>common fraction</u> (with simple integers for its numerator and denominator) that is depicted above.
50	The logs of both of these numbers have the same characteristic of 2, yet different mantissas. The "characteristic" refers to the movement of the decimal point in order to arrive at a number between one and ten.

51	Alex is a circle inscribed in a polygon who looks as if he is very nervous about having no escape and being closed in by all those connecting tangents.
52	Every other real number has a <u>multiplicative inverse</u> except for zero. Division by zero is undefined for real numbers. Thus, the number "0" is lying on a couch talking to his psychiatrist.
53	This exemplifies the first few steps in the <u>method of exhaustion</u> used by the Greeks for areas of geometrical figures ... this was a precursor to Integral Calculus.
54	The <u>group</u> of <u>even</u> numbers is <u>closed</u> under the operation of <u>addition</u>. We have two odd numbers talking to each other and realizing that within the operation of addition (summing), if one adds two members of the set of even numbers together, the result will

	be another even number which is also a member of the set of even numbers. One can never add two even numbers and get a sum which is an odd number.
55	The Stanislavsky School preaches method-acting. Two common Cartesian Coordinates of x and y are discussing the Polar Coordinates above … which are depicted with a Polar Bear on a block of ice!
56	1 is the multiplicative identity … thus, the result of the product of "5 X 1" will clearly be only the "numeral 5"!
57	Troy, the SIGMA notation on the right, where sigma is the Greek upper case letter for "S" for summation, is equal to the expression on the top of the page. The top of the page depicts a more "non Greek" … or a more common way of expressing the result of the integers that are added together from 1 to 5. This is simpler,

	yet this is the "identical twin" summation of these integers which correspond to Troy and his "Greek summation notation".
58	The two <u>definite</u> integrals are discussing the <u>indefinite</u> integral above ... thus the reference to <u>indecision</u>. Also, the indecision can infer that there are also <u>two different</u> <u>substitution choices</u> for solving the indefinite integral at the top of the page.
59	An "<u>anti-log</u>" is an "<u>inverse log</u>" ... and the dictionary definition of "<u>anti</u>" is "<u>opposite or contrary in position</u>". Thus "anti", here, can mean "going against".
60	π is a <u>transcendental</u> number. Mathematically, a transcendental number is a number that is not the root of any integer polynomial. This helped in the proof that "circle squaring" was, in fact, insoluble. Here,

	the two simple integers are talking and comparing this property of being "transcendental" with the English common definition relating the word "transcendental" to the "supernatural".
61	If we consider the equation $Ax^2+Bxy+Cy^2+Dx+Ey+F=0$ with A, B, C not all zero, then, if we look at "B^2-4AC" as the discriminant, then, if the conic section is non–degenerate, then: if $B^2-4AC<0$, the equation represents an ellipse (note: if A=C and B=0, this is a circle, where this is a special case of an ellipse); if $B^2-4AC=0$, the equation represents a parabola; if $B^2-4AC>0$, the equation represents a hyperbola (note: if we also have A+C=0, the equation represents a rectangular hyperbola).
62	Numbers are defined as "<u>amicable</u>"

	numbers when the sum of the proper factors of each one of the numbers adds up to equal the other number.
63	This is a <u>double pun</u> on the word <u>"sign" and the word "sine"</u>. The curve above is the curve **y = -sine x**. Therefore, not only is "sign" and "sine" pronounced the same way... however, we have two "positive sign" sine curves that are discussing a <u>"negative sign" sine curve</u> ... thus of the opposite "sign"!
64	In differential calculus, a "point of inflection" on a curve y = f(x) is a point at which the concavity changes. Concavity upwards or downwards depends upon whether the <u>tangents</u> to the curve are above or below the graph of f(x). Frank, above, is a second derivative that is equal to zero ... this is, in contrast to the second derivatives below that are equal to 1 and -1, respectively. Mathematically,

	it is considered a necessary condition for Frank to be a "point of inflection" to be that f″(x) = 0.
65	In plane projective geometry, one can consider the "Principle of Duality". Here, when the word "point" is replaced by the word "line", and "line" is replaced by "point" … a mathematician can simply prove one related theorem, and by this principle, "the dual of this same theorem" is also valid.
66	These are both two different ways of writing the same "algebraic expressions". The expression that is speaking is an example of earlier "syncopated algebra", while the other is simply common symbolic algebra.
67	Two ordinary derivatives are discussing a directional derivative … thus Pearl knows in which direction she is going.
68	A lemma is defined as "a theorem

	which is <u>used to prove another theorem</u>".
69	Karen is being tickled by the points of the inscribed polygon … her circle is therefore not a strong solid line.
70	A cycloid looks like this curve that frames Steven's room. A cycloid is the curve traced out by the point on the rim of a circular wheel as the wheel rolls along a straight line. The actual curve looks like this … . This capitalizes on the vernacular cry of many parents when referring to their child's messy room …"It looks like a cyclone hit your room".
71	This refers to <u>Zeno's Dichotomy Paradox</u> that for one to be in motion, one must first arrive at the halfway stage before one arrives at their goal. This is called the Dichotomy Paradox because it involves repeatedly splitting a distance into two parts.

	Thus, by continuously halving the distance to the finish line, one can never reach the end ... because there will always be one more distance remaining to be halved and no one will ever reach the line at the end of the race track.
72	This is the "Witch of Agnesi" curve in analytic geometry ... witches are most commonly associated with Halloween!
73	This refers to Gauss' "Rule of Nine". Here, we see that if the difference of two numbers results in a number divisible by 9, then, after observation, we see that the two original numbers had their digits interchanged.
74	The height of an equation is a mathematical term that is evaluated by "adding all the sums of the absolute values of the coefficients together, and then adding to that value the degree of the equation minus one" ... this is, by definition,

	equal to the "height of an equation".
75	Two regular pyramids are discussing a "<u>truncated</u>" pyramid. A "truncated" pyramid is one that is cut by a plane parallel to the base and has the apical part removed.
76	Two differential equations are discussing corresponding primitive equations.
77	When the second derivative of a function is positive, this means that there is a minimum point. However, with the equation that is talking, the second derivative will always be negative, thus implying a maximum point for this equation. Thus, for the talking equation there will only be a maximum point for any value of x where this first derivative is equal to zero … thus, implying no slump, just a maximum. However, the above equation, similarly, always has a positive second derivative for any

	value of x where the first derivative is zero ...thus, indicating a minimum value that slumps to a downward point.
78	Harold is a <u>Hermitian</u> matrix. This is alluding to the similarity of the word Hermitian and a "hermit". A hermit, in English, is usually considered to be a person who lives, to some degree, in seclusion from society.
79	This is the equation of the <u>growth curve</u>. It is most often used in relation to bacteria and their normal reproduction.
80	In algebraic braid theory this is the representation of the common <u>hair braid</u>. Braid Theory was invented in the 1900's by Emil Artin, and is most commonly considered as a branch of Algebraic Topology.
81	This exemplifies the <u>Law of Inference</u> in the field of Mathematical Logic.
82	This refers to one of the three most

	famous geometric problems unsolvable by compass and straight edge construction … the "Doubling of the Cube". "The duplication of the cube" requires one to construct a new cube, with Euclidean tools, whose volume is two times the volume of the original cube. This is also sometimes referred to as the Delian Problem. This problem is often believed to be originally related to "ending a specific plague by doubling the size of an altar to Apollo".
83	An "ex-circle" is a mathematical term used when a circle is tangent to one side of a triangle and this circle is also tangent to the extension of the other two sides of this same triangle.
84	This is the "Sieve of Eratosthenes" for prime numbers. Although discovered in 240 B.C., this Sieve is still considered to be a very efficient method for finding all prime numbers

	less than a specific integer. Here, the word "Sieve" can also be viewed as something used in the kitchen for straining solids from liquids when preparing culinary delights.
85	When one <u>operates</u> on a matrix by <u>augmenting</u> it … the column of y's only represents the <u>identity</u> matrix. Here, the matrix that is talking still has the identical coefficient entries as the above augmented matrix.
86	This refers to the <u>transformation of area</u> by the use of the mathematical fact that … as long as the base remains the same and the altitude is equal, the new triangle has the same area as the old one. Here, we see that by a succession of these operations, and by several iterations of this concept, one can construct various new bases within the original figure in order to arrive at the final "shapely version" of Emma.

87	The equation given is the one for a "four leaf clover" in Analytic Geometry. Also, the number "13", that is talking, is often considered to be the antithesis of "good luck"!
88	Here the identity matrix is talking to the officer about a different use of the word "identity" … alluding to the illegality of "identity theft".
89	What is pictured here is the "cardinal approach". The opposite method, mathematically, is the "ordinal approach". However, by capitalizing on the reference to the Roman Catholic Church's Cardinals … perhaps the opposite might be the "sacrilegious approach".
90	The Hindu's always found the sine of a "Half-Chord". Perhaps, a guitar-based folk song could be played, not in the chord of "C", but in the chord of "C/2". Here we parallel the two definitions of the word "chord" … one

	is mathematical and the other is musical.
91	The Pythagoreans considered <u>odd numbers</u> to be masculine and even numbers to be feminine. Here, the numbers 4 and 5 represent the Mom and the Dad. They are added together to give birth to the number 9 … who, by similar consistency, would also be a male.
92	A X B = $\|A\|\|B\|\vec{n}\sin\theta$ … one should also note that when one breeds animals or plants, one often refers to, and utilizes, intentional "crossbreeding" in order to create offsprings that share the traits of both parents' lineages. Here, the Monkey and the Lion are capitalizing on both the "vector mathematical" and the "biological" interpretation of the "cross product"–ing of two entities.
93	Poisson Distribution is a mathematical term, which is commonly depicted by

	the shape of this graph. However, in French, "Poisson" also means "fish". In probability theory, the Poisson Distribution expresses the probability of a given number of events occurring in a fixed interval of time if these events occur with a known average rate and independently of the time since the last event.
94	The Fibonacci Sequence was developed and is often used in reference to the description of the rate of rabbits' breeding in ideal circumstances. Thus, the Fibonacci numbers actually represent the number of pairs of rabbits as they reproduce each month.
95	$r^2 = a^2\cos2\theta$ is the equation of the lemniscate curve which happens to resemble the symbol for infinity.
96	The two CATenary curves are talking about another figure which is made up of eight of these CATenary curves.

	The term octopus comes from: "octo = eight", and the "pus" comes from "pussy cat" that = <u>CAT</u>enary curve". Also, the catenary, whose equation is given, is the curve obtained by <u>hanging</u> a cable, or a clothes-line, from two poles or perhaps two nails, etc. ... thus the reference to "hanging around" with the other two curves!
97	These surfaces are all <u>topologically equivalent</u>. However, in order for the flower pot to be surface equivalent to the tea-cup, the <u>flower pot</u> must have a <u>hole in the bottom</u>. With this hole, it can't hold water very well. Here there is a play on the idiomatic phrase "does not hold water", which also implies that "something does not seem to be true or reasonable" ... such as the doughnut, coffee cup and flower pot all being topologically equivalent ... especially as it relates to a "non-super-topologically-friendly"

	observer!
98	A conic section can be defined as the locus of points whose distance from a given focus, divided by the distance to a given line called a directrix, equals a constant ... this is defined as the eccentricity of the conic section. For an ellipse this ratio is greater than 0 and less than 1. For a parabola, this ratio is equal to 1. For a hyperbola, it is greater than 1, and for a circle, the eccentricity is equal to 1. Here, Davis, at the top of the page, has eccentricity > 1 ...while the talking conic section, which is an ellipse, has an eccentricity greater than 0 and <1, and the parabola's eccentricity is equal to 1!
99	These are both examples of the equation for <u>Damped Harmonic Motion</u>. Often one thinks of Damped Harmonic Motion in reference to springs, swinging pendulums, or

	vibrating violin strings. Here, there is also a comparison of the English word "damp" to the atmospheric condition of "humidity"!
100	Isometric, in mathematics, means being equidistant from each of three axes. However, the common use of the same word, relates to a method of exercise for figure-molding and improving one's physical appearance. Isometrics is a type of strength training in which the joint angle and muscle length do not change during the contraction exercise. Isometrics are done in static positions, rather than being dynamic and through a range of motion.